CASHEWS

by Emma Bassier

Cody Koala
An Imprint of Pop!
popbooksonline.com

abdobooks.com
Published by Pop!, a division of ABDO, PO Box 398166, Minneapolis, Minnesota 55439.

Printed in the United States of America, North Mankato, Minnesota.

052021
092021

THIS BOOK CONTAINS RECYCLED MATERIALS

Cover Photos: iStockphoto, foreground; Shutterstock Images, background
Interior Images: iStockphoto, 1 (foreground), 7 (bottom left), 7 (bottom right), 8, 21 (top), 21 (bottom left), 21 (bottom right); Shutterstock Images, 1 (background), 14, 17; Alan Skyrme/Alamy, 5; Red Line Editorial, 7 (top); Jutta Ulmer/mauritius images GmbH/Alamy, 11; Jose Mathew/Alamy, 13; Sylvia Buchholz/Reuters/Alamy, 15; Dinodia Photos/Alamy, 18

Editor: Aubrey Zalewski
Series Designers: Laura Graphenteen and Colleen McLaren

Library of Congress Control Number: 2020948851
Publisher's Cataloging-in-Publication Data
Names: Bassier, Emma, author.
Title: Cashews / by Emma Bassier
Description: Minneapolis, Minnesota : Pop!, 2022 | Series: How foods grow | Includes online resources and index.
Identifiers: ISBN 9781532169779 (lib. bdg.) | ISBN 9781098240707 (ebook)
Subjects: LCSH: Cashew nut--Juvenile literature. | Nut industry--Juvenile literature. | Nut trees--Juvenile literature. | Agriculture--Juvenile literature. | Food crops--Juvenile literature.
Classification: DDC 631.5--dc23

Hello! My name is

Cody Koala

Pop open this book and you'll find QR codes like this one, loaded with information, so you can learn even more!

Scan this code* and others like it while you read, or visit the website below to make this book pop.

popbooksonline.com/cashews

*Scanning QR codes requires a web-enabled smart device with a QR code reader app and a camera.

Table of Contents

Chapter 1

Curved Seeds

Cashews are curved, tan seeds. Plants **reproduce** with their seeds. Cashews grow on trees. The trees stay green all year long. They are shaped like umbrellas.

Watch a video here!

Chapter 2

Growing Cashews

Cashews are **native** to Brazil. They grow in **tropical** places. Cashews are most common in India and Vietnam. They also grow in some African countries.

Where Cashews Grow

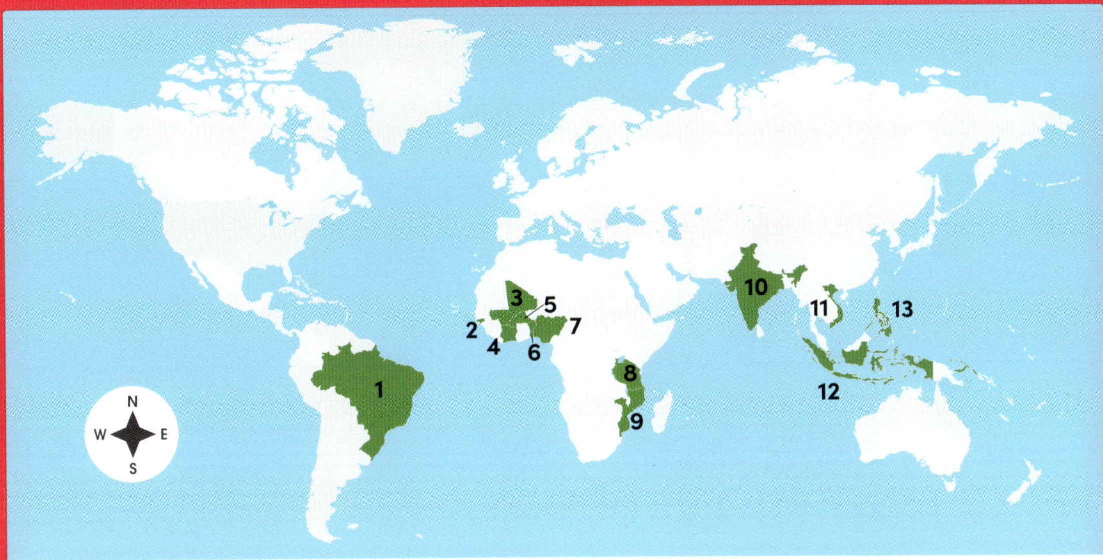

1. Brazil
2. Guinea-Bissau
3. Mali
4. Côte d'Ivoire
5. Burkina Faso
6. Benin
7. Nigeria
8. Tanzania
9. Mozambique
10. India
11. Vietnam
12. Indonesia
13. Philippines

Complete an activity here!

cashew apple
cashew

Cashews grow off the end of a **swollen** stem. This stem is called a cashew apple. It is usually yellow or red. The apple tastes sweet and sour.

Cashew apples look like fruit. But they are false fruits. They grow from parts of the plant other than the flowers.

Cashews have two shells. The outer shell is thin and smooth. It turns from green to brown as the cashew ripens. The inner shell is very hard.

outer shell
inner shell

Chapter 3

Preparing Cashews

It takes approximately two months for cashews to grow. The cashews grow during the winter or the dry season.

Learn more here!

Cashew apples fall to the ground when they are ripe.

Farmers also pick the apples by hand. Then they twist off the cashews.

People can eat cashew apples. Sometimes people make juices and jellies with them.

A sticky **resin** lies between a cashew's two shells. It is **toxic**. It can hurt people's skin. The cashews must be **roasted**. The heat causes the outer shell to burst open. The heat burns away the resin.

Workers then break open the inner shells. They use a machine that presses down on the shells. Next, workers dry the seeds. Then they carefully peel off the seeds' papery coverings.

Workers can shell approximately 22 pounds (10 kg) of cashews a day.

Chapter 4

Eating Cashews

Cashews are common in Southeast Asian and Indian meals. People also enjoy eating cashews as a snack. Cashews have a sweet, nutty flavor.

Learn more here!

Making Connections

Text-to-Self

Did you know that cashews grow with cashew apples? What other kinds of foods grow in surprising ways?

Text-to-Text

Have you read other books about how foods grow? How were those foods similar to or different from cashews?

Text-to-World

Cashews can be used in meals or eaten as a snack. What other foods can be both meals and snacks?

Glossary

native – naturally living in a certain area.

reproduce – to make new offspring.

resin – a thick liquid that comes from plants.

roast – to cook in a dry heat in an oven or over a fire.

swollen – describing something that has grown larger than usual.

toxic – poisonous.

tropical – having to do with a place where the weather is usually warm and wet.

Index

Online Resources

popbooksonline.com

Thanks for reading this Cody Koala book!

Scan this code* and others like it in this book, or visit the website below to make this book pop!

popbooksonline.com/cashews

*Scanning QR codes requires a web-enabled smart device with a QR code reader app and a camera.